Trucks of the Eighties

Colin Wright

This red and white liveried Leyland 'Road-train' registered F945 JCH was fitted with aerodynamic fairings and when photographed was coupled to a tri-axle trailer. It was operated by Machine Mart – the tool and machinery superstore group, this vehicle working out of the Alfreton distributing centre. *Colin Wright*

CHANNEL VIEW PUBLICATIONS
Clevedon · Philadelphia · Adelaide

Winter sun, spray and slush all combine to enhance this view of a MAN diesel operated by R. F. & S. A. Hewetson, registered B423 HAO. This wintry view was taken on the Accrington by-pass and shows a couple of tri-axle semi-trailers on a third similar unit, the two being en-route to their new owners from their manufacturer – SDC in Northern Ireland. *Colin Wright*

EDITORIAL

This volume of *Trucks in the Eighties* is the first in a series planned to cover a wide range of commercial vehicle themes. My aim is to present interesting photographs with reasonably detailed captions, a policy which in the past has worked well judging from the amount of interest and correspondence generated by the *Trucks in Britain* series. *Trucks in the Eighties* is laid out for the most part utilising double page spreads based on a 'theme' – Wreckers, Supermarket Transport, Removals, etc. Next in this series will be a volume covering the eight wheeler's along with one devoted to council vehicles – a so far neglected area, but one full of interest. I hope this volume gives as much pleasure to the reader as it has to me whilst preparing it. To the drivers and operators who have been photographed but do not see their vehicles included – my apologies and better luck next time!

I would like to express my thanks to many friends and companies for the enormous amount of help given. A number of friends have taken photographs for me and those are all credited in the respective sections. Special thanks are due to Geoff Longbottom and Bob Tuck for photographs and for supplying information.

Colin Wright,
Damside Cottage, Kettlewell,
Nr Skipton, North Yorkshire. *1/4/94*

Library of Congress Cataloging in Publication Data
Wright, Colin, 1939–
Trucks of the Eighties / Colin Wright
1. Trucks–Pictorial works. I. Title
TL230.12.W75 1994
629.224–dc 20 94–26003

British Library Cataloguing in Publication Data
A CIP catalogue record for this book is available from the British Library

ISBN (pbk) 1-873150-03-2

Channel View Publications
An imprint of Multilingual Matters Ltd
UK: Frankfurt Lodge, Clevedon Hall, Victoria Road, Clevedon, Avon, England BS21 7SJ.
USA: 1900 Frost Road, Suite 101, Bristol, PA 19007, USA.
Australia: PO Box 6025, 83 Gilles Street, Adelaide, SA 5000, Australia.

Typeset by Wayside Books, Clevedon, Avon. Printed and bound in Great Britain by Amadeus Press Ltd, Huddersfield.

EIGHT WHEELERS

F315 DKM and E982 TKK were Volvo 'FL7s' operated by Brett, of Canterbury – whose tipper fleet was well known south of the Thames. They were photographed in home territory near Folkestone. *Colin Routh*

The fleet name on F984 BKP pleasingly uses an old English type-face. I. W. & K. F. Messenger & Sons were the operators of this Leyland 'Constructor 30-26' which sported the narrow cab – one used infrequently on eight-wheelers. The location of this view is again near Folkestone. *Colin Routh*

WAGON AND DRAG

When built, this Mercedes Benz '1644' (one of the first to be fitted with EPS) registered D483 KWT was one of the largest curtain-siders in service. Bodywork was by Wilsons Truck Services of Bingley and had a capacity of 48m^3, and the trailer 67m^3. Its operator was Brian Yeardley, based at Featherstone (West Yorkshire). *Colin Routh*

C939 GBH was an ERF 'C' series in the livery of Myer's beds. It was recorded approaching Barton lorry park in County Durham. *Colin Wright*

In contrast to the previous view E768 OCL had a demountable body system by Able Systems. This Volvo 'F10' was operated by Murfitts of Ely and was photographed on the A1 at Norman Cross. Bodywork of both lorry and trailer was by Crane Fruehauf with a combined capacity of some 110m^3.

Colin Wright

Another demount system is featured in this view of Mercedes Benz '1625' registered E715 BDM operated by East Lancastrian haulier T. & T. Thompson Ltd on contract to Kellogg's. Bodywork is by Cartwright and the trailers fitted with a Pietz coupling, allowing truck to trailer distance to vary according to the tightness of the turn.

Colin Wright

PARCEL CARRIERS

The business of parcel carrying has increased enormously over the past few years with a number of large operators handling most of the traffic.

The first of our four 1980s photographs shows Independent Express M·A·N '20-331' registered D375 YRP, The company employed a pink livery for its vehicles; this one was photographed in Wigan. Independent Express are no longer in business having suffered a take-over.

Colin Wright

Still very much in business N.F.C. Lynx use a black livery. This Volvo 'FL7' registered D26 KND was one of a large fleet of Volvo's operated by the company, this view being recorded near Wigan.

Colin Wright

In evidence during the 1980s but now no longer in parcels traffic, Federal Express used an attractive red, white and blue livery, here carried by F614 SEE, a Mercedes '1625' hauling a Carrymaster tandem axle semi-trailer. *Colin Wright*

The white and orange livery of TNT is recorded in this photograph of F713 JBA, an Iveco Ford tractor unit seen hauling a Cartwright-built semi-trailer, its sides proudly proclaiming 'Haulier of the Year'. The fleet no. 2713 is displayed on the cab front and is a give-away means of identification of TNT vehicles on contract to other companies.

Colin Wright

PIGGYBACK OPERATIONS

A quartet of 'Piggyback' operations!

Seddon Atkinson '400' series registered BEF 133T was photographed approaching the A1 near Ripon (North Yorkshire) having travelled down from the north-east. The operator was Youngs, based at Stokesley. The load – one of their Volvo 'F7s' on a single axle Beaver-tail semi-trailer.

Colin Wright

DCW 590Y was a newer Seddon Atkinson – this time a '401' powered by the ubiquitous Gardner engine. No. 70 in the fleet of W. & J. Riding (of Longridge, near Preston, Lancs), this blue and grey liveried rig on which sister vehicle No. 86 shared the load space with a sheeted part load. The location was Halton East on the hilly A59 on the way to Skipton from the north-east.

Colin Wright

HWG 610W, a Scania '81' from the early 1980s, was operated by Cramscene Ltd, of Morley near Leeds and was photographed at Norman Cross on the A1 near Peterborough. The load was a six-wheel bonneted Magirus Deutz tanker. *Colin Wright*

B931 UKH, a Leyland '17-28 Roadtrain' was operated by G. W. Sissons & Son Ltd and its load was one of their Scammell 'Crusader' tractor units. It was seen outside the service area at the intersection of the A1 and M62 at Ferrybridge. *Colin Wright*

HOUSEHOLD/THEATRICAL REMOVALS

Household removals have been associated with the transport business since the horse-drawn era. However, there is nothing old-fashioned about the business today with the latest vehicles being employed. Of the four views reproduced, three are true household removal vehicles whilst the fourth is associated with the entertainments industry.

Willis of Skipton operated A34 HPB, a Bedford 'TL' with an integral body by Vanplan having a capacity of about 2,300 cubic feet and plated at 15 tonnes (although removal vehicles generally run quite a bit lower than their plated weight). A34 HPB was photographed at the author's business premises where it had parked to off-load furniture to a smaller vehicle for delivery to a near-by house where access was restricted. *Colin Wright*

The second photograph in this sequence shows GB Liners C468 XDF based at Hereford. This vehicle looks more traditional with the standard Leyland cab whilst the Marsden bodywork incorporates a cab-top sleeping berth. It was seen in the Cotswold village of Broadway during 1989 at a time when cobbling of the roadways round the village green was in progress. *Colin Wright*

Northern Ireland based KOI 187O carried a Vanplan body on a Leyland 'Leopard' p.s.v. chassis. Operated by the Dixon Line, it, like the other vehicles illustrated, uses the van sides to good advantage to advertise their businesses. A long way from home KOI 1870 was photographed during 1984 at Threshfield in the Yorkshire Dales. *Colin Wright*

The final vehicle in this block of four does not quite fit into the household removals bracket, but nevertheless is used on similar duties as a theatrical removal van. The livery of the DAF '2100' reflects its role in life – note the elaborate paint job complete with theatrical masks on the cab door. Note also the crew quarters behind the cab. The operators of C140 FGT were G. H. Lucking & Sons of Washington, Sussex. *Phil Sposito*

W. P. Twibell & Sons Ltd (of Mobberley, Cheshire) were the operators of ABU 716Y, a Dodge '300' series tractor. These units were in fact produced in Spain by Barreiros who, like Dodge, were a subsidiary of the Chrysler Corporation. The tandem axle trailer had a Boalloy body on Crane Fruehauf running gear. *Colin Wright*

The blue and white livery of the Ford Motor Company was carried by C863 LVX, one of that company's 'Cargo' models hauling a King & Taylor body and in use as inter-factory transport. It was seen at Ross-on-Wye about to join the M50 in 1989. *Colin Wright*

With the lamp post being perhaps the focal point of this photograph, a pair of T. B. Oliver Ltd's Mercedes Benz tractor units were seen approaching the A1 at Dishforth. Leading is a tri-axle model whilst behind is a two-axle unit. Both are coupled to trailers built by Lawrence David which were used primarily for transporting produce. *Colin Wright*

Recorded on the A40 at Abergavenny, D631 AVJ operated by Oakley, Hereford was a Seddon Atkinson '401', here coupled to tri-axle trailer unit built by Boalloy. The electricity pylon seen in the photograph is one of a number that dominate the A465 trunk road in this area. *Colin Wright*

CAR TRANSPORT

Car delivery has changed a lot since the days when each vehicle was driven from the factory to the dealers by drivers who would hitch-hike back home carrying a pair of trade-plates.

A Mercedes '1626' provided the power for this small transporter built by Alastair Carter (of Tamworth). Registered RKR 357W this unit was operated by Richard Lawson (of Kirriemuir). These semi-trailers were renowned for their simple and light build.

Phil Sposito

Grayston Transport operated this DAF '2500' registered A246 VCA seen travelling north about to join the M50 at Ross-on-Wye. Note the position of the cars on the bottom deck of this Mk2 Hoynor, modified to accommodate greater loads by the addition of flip-up ramps.

Colin Wright

Carfax were the operators of B423 JFU – a Renault Turbo, hauling a Hoynor Mk7 trailer with a modification known as the hammock, designed to increase the capacity to eight medium-sized saloons. The eighth car can just be seen below the third and fourth cars on the top deck. *Colin Wright*

One of the country's leading operators of car transport was Abbey Hill from Yeovil – the company name being R. K. Bastable Ltd. The trailer unit in this instance was an Iveco hauling a triple-deck Hoynor semi-trailer with a capacity of 10 cars. When photographed only six cars were aboard as the driver was part way through his deliveries in Warrington. *Colin Wright*

BREWERY WAGONS

In 1985/6 Joshua Tetley & Son, the Leeds-based brewers, carried out a vehicle evaluation programme to formulate future buying policy when replacement of the then 10 current fleet of Ford 'D' series 'Chinese Sixes' became necessary. The vehicles used in this assessment were a Renault/Dodge, Leyland 'Freighter' and a Ford 'Cargo' – all curtain-siders, and an ERF with the more standard dray bodywork. The first three types were designed to haul a two-axle trailer which could be parked at a suitable point when empty to await collection by the returning truck later in the day. It was subsequently decided to order Ford 'Cargo's' fitted with York third axles. The four vehicles tried out were:

Fleet no. JTG3, a Renault/Dodge registered C52 CUA, delivered 10 March 1986 and photographed outside the Clarendon Hotel at Hebden – one of the excellent hostelries in Upper Wharfedale. *Colin Wright*

Fleet no. JTG2 was a Leyland 'Freighter' registered C681 AYG, delivered 8 November 1985. The trailer hauled by this vehicle was made by Wheelbase Engineering, Blackburn. It was photographed at the Sheffield premises at Joshua Tetley & Son. *Colin Wright*

Our third photograph illustrates JTG4 – Ford 'Cargo' C555 OCP, delivered 12 June 1986 and photographed in Bridlington.

D. G. Savage

The ERF 'M16' C960 GJX carried fleet no. JTG1 (JTG = Joshua Tetley Goods) and was delivered on 1 November 1985. It was seen in Joshua Tetley's Leeds premises.

Colin Wright

REFRIGERATED TRANSPORT

Refrigerated vehicles were a common sight on the roads of the 1980s and generally were kept in a clean condition as befitted their role as food carriers.

Most Little Chef restaurants were served by Puritan Maid delivery vehicles and this Volvo 'F6S' was seen during the driver's lunch-break at the Hartshead Moor Services on the Trans-Pennine M62. B250 YYA carried bodywork by Grey & Adams, of Fraserburgh, with a Petter 'Loc 4' fridge unit.

Colin Wright

The M62 again features in our second view, this time showing a Birds Eye/Walls Leyland 'Freighter' registered C301 7AD. This vehicle was one of a growing number of 'Chinese Six' versions of the standard four-wheeler becoming increasingly popular in the late 1980s.

Colin Wright

Seamer Transport Ltd operated this smart Foden A687 JWU. It was photographed near Warrington having just delivered its load to the Safeway Distribution Centre. The sides of the York semi-trailer (fitted with a Petter 'PDL50' fridge unit) have been effectively used to advertise McCain Beefeater chips. *Colin Wright*

West Midlands based Coldlink's ERF 'E10' registered F508 YFD was seen at Norman Cross on the A1. A Thermo King 'SB250SLE' fridge unit was fitted to the Schmitz Ferroplast bodywork. *Colin Wright*

SUPERMARKET TRANSPORT

In contrast to the household removals business, supermarkets are a fairly recent innovation starting to move into full swing in the 1950s. Four of the best known (in the North) are featured here (with apologies to Sainsbury's). The standard of all the fleets is high as befits the food trade.

Ferrybridge Power Station in West Yorkshire makes an imposing background to F283 LCA. Numbered S17 (S = Sherburn-in-Elmet) in the Kwik Save fleet, this was one of a large number of DAFs, a type favoured by this well-known discount store. In this instance motive power is a '2500' hauling a two-axle box semi-trailer. *Colin Wright*

C888 MVH was painted in a lined black livery which is striking when coupled to light-coloured trailers. The main centre panel was painted red, green, blue or yellow and the slab sides were used to advantage to advertise Morrisons' products. No. 55 in the fleet, the tractor unit was a MAN '16-291' and the only one then in the fleet fitted with a sleeper cab. The semi-trailer was by York which whilst plated at 38 tonnes, would of course not exceed 32 tonnes with the two-axle trailer. Morrisons operational area stretches from the Scottish border, south to Lincolnshire. *Colin Wright*

E721 PWY was a Seddon Atkinson '4-11' operated by Asda (head office in Leeds). The all-over white livery was effective, being devoid of any unnecessary embellishments – simply stating boldly the company name. The trailer, plated at 32 tonnes, was built by Saxon and had an under-slung freezer unit by T. F. Refrigeration, Gosport, this latter feature contributed to the clean lines of the whole rig which was photographed outside the company bacon plant at Lofthouse (near Leeds). Asda vehicles can be seen from Cornwall to Scotland and can claim to operate nationwide. *Colin Wright*

A Mercedes '1625' tractor unit registered E425 NDC was photographed hauling a York-built semi-trailer in Safeway livery with the more traditionally positioned Thermoking refrigeration unit. It was seen approaching the distribution centre on the outskirts of Warrington in the summer of 1989. *Colin Wright*

GIGANTIC PACKAGING

Who would like a Kit-Kat bar this size? These views of Rowntree distribution vehicles are mainly of interest for the liveries carried on their trailers which are instantly recognisable – who indeed has not seen the 'Yorkie Bar' advertisements on television? All four tractor units illustrated were supplied through Reliance at Brighouse and all were 6 x 4s designed to operate at maximum 32.5 tonne weights.

Our first view shows C974 YWY (fleet no. 5031), a 'C' series ERF, approaching the M6 near to Rowntree's Warrington facility. The 40ft box trailer, built by Crane Fruehauf, was in the red and white colour used for Kit Kat chocolate labels. *Colin Wright*

Carrying on the theme of reproduction of product wrappers, this liquid chocolate tanker in Rolo livery, was being hauled by C50 BUA (fleet no. 5050) – another 'C' series ERF. The Thompson-built tank was insulated with electric blanket heating to keep the chocolate in molten state whilst being transported from the factory at York to satellite factories. *Colin Wright*

C734 AUG carried fleet no. 5036 and was a third 'C' series ERF, all of which were powered by Rolls Royce '340' engines with Spicer gearboxes. Probably the best known of all the Rowntree liveries was the basically blue Yorkie Bar, photographed at the Halifax factory in 1989. The cab roof deflector was built in-house at York as were the skirt panels on the Sun-Pat trailer seen in our last photograph.
Colin Wright

Rowntree's were devotees of the ERF and F596 AWW (fleet no. 5072)) was an 'E14' powered by a Cummins engine; the trailer carrying the attractive Sun-Pat Peanut Butter livery.
Colin Wright

PETROL AND OIL PRODUCTS

3YE1270 is just one of the Hazchem codes, in this instance applied to petroleum products.

Red Texaco liveried D202 GYG was operated on contract from Tankfreight. This Volvo 'F10' tractor unit which had been supplied by Crossroads Commercials, was coupled to a tank built by Gloster Saro and the rig was photographed en-route to the M1 in Leeds. *Colin Wright*

GHX 582W was one of a large fleet of yellow and white livery vehicles operated by Shell. This ERF 'B' series was photographed passing the Portman Building Society on the edge of Worcester's pedestrian precinct. *Colin Wright*

A brace of Esso tankers photographed within seconds of the Texaco one (first photograph). In the lead F180 POU, a Seddon Atkinson '3-11', behind, B104 WYY, another Seddon Atkinson, but this time a '401' unit. *Colin Wright*

Leyland/DAF '95' series F959 THD in the red and white Total livery, was hauling an un-lettered semi-trailer when photographed at Hartshead Moor Services on the M62. *Colin Wright*

TIMBER HAULAGE

The photographs on these two pages show various stages in the processing of timber ranging from trees out of the forest, to finished products.

First we see Waddington's Bingley (West Yorkshire) based Foden registered D353 GWX approaching its home base at Crossflatts with a load of round timber en-route to the saw-mill for processing; in this case most likely as chock wood for British Coal.

Colin Wright

Scottish-based James Baxter & Son (from Pathhead, near Dalkeith) DAF '3600' registered C722 SSF is seen a long way from home, approaching the A1 east of Thirsk. The load in this case is pulp wood destined for the pulp mills for paper making. *Colin Wright*

A Japanese Hino 'SH28' was photographed at Ross-on-Wye in the summer of 1989 carrying a load of imported sawn timber, probably for pallet wood. F640 YON was operated by Weirfleet Services, of Coventry. *Colin Wright*

A part load of roof trusses for use in one of the many barn conversions now taking place, was seen on a tandem-axle semi-trailer with singles. Motive power was provided by Leyland 'Cruiser 16-21' A146 MEW. The operators – Harrison & Lewin Ltd – were based at Barton (Lincolnshire) but this photograph was taken some 150 miles from home at Conistone-with-Kilnsey in the heart of the Yorkshire Dales. *Colin Wright*

COAL HAULAGE

During the 1980s a considerable tonnage of coal was carried by road and these four photographs show coal being transported from colliery to Buildwas Power Station in Shropshire. All were taken in the winter of 1988/9 before completion of the Ironbridge by-pass. Before the opening of the new by-pass coal lorries had to negotiate Jiggers Bank but this route was liable to subsidence problems.

Our first photograph depicts Leyland 'Constructor' C477 AAE in the blue livery of Hughes Transport, Ironbridge. The premises of this company were near the top of Jiggers Bank. This eight-wheeler was seen about to climb up the Bank. The sign, prominent in this view, advertises the world renowned site of the birth of the Industrial Revolution. *Colin Wright*

PYG 331M seen in the second photograph is a Scammell 'Routeman' eight-wheeler operated by Dave Beaman and recorded at the top of the Bank on its way to pick up another load. *Colin Wright*

Clay Colliery Co. Ltd Scottish registered Volvo 'F7' eight-wheeler B937 YHS at Dale End at the Ironbridge end of Jiggers Bank. *M. Kent*

Mercedes Benz '3025' C787 NRR was seen heading a convoy of trucks towards Buildwas. This eight-wheeler was photographed in its demonstrator livery prior to being painted in the attractive blue livery of J. H. Parry & Sons (Shawbury) Ltd. *M. Kent*

CEMENT MIXERS

Cement mixers on eight-wheel chassis are fairly rare, the one illustrated being operated by a small family business, whilst the three six-wheelers are from national companies.

AJV 133W, a Seddon Atkinson, was fitted with a Ritemixer body built by D. Wickham, Ware, Herts. The operators – R. W. Potter (Boroughbridge) Ltd – are sand and gravel merchants in this small North Yorkshire town once bisected by the A1 until a by-pass came into use. Potter's fleet comprised mainly tippers sporting a dark blue livery. *Colin Wright*

JUA 945V (fleet no. 4766) in the Tilcon Trumix maroon and blue livery dated back to 1979 and carried the long used Foden 'S39' cab that had its production run extended primarily for mixer work. *Colin Wright*

Hinckley Island Hotel on the A5 was the location in which Leyland 'Constructor 24-17' registered E659 BBW (fleet no. 4582) was photographed. The mixer was operated by Redland Aggregates. *Colin Wright*

In green and white livery, Pioneer were the operators of Mercedes Benz '2421', registered E161 PWO (fleet no. 7059). The mixer was by Ritemixer and the location was Erith, Kent, in 1989. *D. G. Savage*

MILK TRANSPORT

Originally, milk was collected from the farms in churns and previous to the mid-1950s collection was entirely by this method. Churns finally went out in August 1978 and all farm collections are now made by tank wagons.

Above: C261 RPL shows a new breed of farm collection vehicle. This Leyland 'Cruiser TL' carries a refurbished 9,000 litre capacity Wincanton tank whilst its trailer has a similar tank taken from an earlier Leyland four-wheeler, but stretched to increase its capacity by some 2,000 litres. Farm roads in general would not be suitable for an outfit of this type and the trailer is therefore left at some suitable point whilst the 'mother' tanker collects the milk and then transfers it into the trailer before going off again for another load. *Colin Wright*

Below: This photograph depicts another aspect of transporting milk. Two four-wheel rigid Dodge vehicles are seen 'feeding' two artics. The two Dodges (together with another pair of four-wheelers) collected milk from local farms in Wensleydale, whilst the artics were used to trunk the milk to the dairy at Blaydon near Newcastle-upon-Tyne. The left-hand artic registered E428 FPF was a light-weight Foden and one of a pair operated. Its trailer carried 25,000 litres and was manufactured by Butterfield. Behind the Foden is B474 NPA – an ERF 'C' series with a 21,000 litre tank. The location of this view is between Leyburn and Richmond. *Colin Wright*

Two small artics registered F841 RPB and F842 RPB were used in a vehicle evaluation exercise for farm collection vehicles – a further move in the on-going evolution of milk transport. F841 RPB was a Ford 'Cargo' and one of two such vehicles, both of which were based at Northallerton. F842 RPB – a DAF '2100' was photographed near Preston (Lancashire) and proved to be the type favoured after the evaluation. Both these trailers were hauling 14,000 litre capacity Clayton/Dairy Products-built tanks.

Colin Wright

RAIL RELATED

One of the largest sources for railway locomotives for transport by road in recent times must surely be the Barry (South Wales) scrap yard of Woodham Bros. Over the years in excess of 200 locomotives have been rescued from Barry for use on the preserved railways scattered throughout the country, and well over half of these moved by road (earlier rescues were transported by rail).

BR Standard Class 4 2–6–0 No. 75014 was photographed on the M62 in West Yorkshire bound for the North York Moors Railway in 1987. The tractor – JPY 958W, Scania '145' operated by A. V. Dawson Ltd, of Middlesborough – was hauling its load on a Nicolas trailer. This machine was orginally built as a dump truck.

Colin Wright

RWO 73R was one of the well-known fleet of Wynns (Newport) – one of 'the' heavy haulage operators. Named 'Superior' this Mark II Scammell 'Contractor' was being very much under used in the haulage of this (to them) very small load. The 'small' load was an ex-Southern Railway Bulleid 4–6–2 rebuilt by British Railways. It was being transported from the North York Moors Railway to the Severn Valley after partial restoration. This 'West County' had languished in Barry for 15 years and was No. 34027 'Taw Valley'. It was photographed at Keele Services.

P. Lee

Somerset haulier Mike Lawrence was a regular visitor to Woodhams. The American connection shows UYC 39W, a Mack 'Interstater'; rated at 145 tons gross. It was seen hauling No. 34028 'Eddystone', another rebuilt 'West County' – on a Transquip trailer, specially equipped with 40ft of railway track along its deck. *P. Lee*

A more mundane load was being carried by this Scania registered SNV 168H, also operated by Mike Lawrence, of Burnham-on-Sea. The Stanier tender was for 'Black Five' Class 4–6–0 No. 45163 and was about to depart from Barry, for Hull Dairycotes. *P. Lee*

AMERICAN TRUCKS ON BRITISH ROADS

American manufactured trucks have never been popular on UK roads and the sight of an American truck is something of an event. The American description of cab types differs from ours – a normal control in the UK is a conventional across the water, whilst our forward control is to the American, a cab-over. In this series of photographs we illustrate two of each.

YWE 784T was a Mack 'R685 RT' operated by G. Preston Jnr of Earby, Yorkshire (Earby was formerly in Yorkshire but following local government reorganisation is now in Lancashire). This vehicle started life with R. J. Norman of Barnsley as a tractor unit usually having a tipper. It was 'stretched' to 11 metres by Wheelbase Engineering of Blackburn and fitted with the curtain-side body pictured. The Mack carried the name 'Midnight Mist'. *Colin Wright*

C402 XDX was even more of a rarity. Built in British Columbia in 1985 this Western Star was the pride and joy of its owner Geoff Byford from Barrow in Suffolk. Geoff purchased the rig from a dealer in Dallas (Texas) when it was some two years old and has since attended numerous shows and rallies in this country. The vehicle has also featured on a poster, the proceeds of which were donated to the Great Ormond Street Hospital. C402 XDX earned the name 'Lady Louise' in memory of Geoff's daughter. *D. G. Savage*

Kenworth cab-over tractor unit registered FHT 332W was operated by Whitwills and it was photographed near to its Avonmouth home. Note the 'smokestack'. *P. Sposito*

This White was registered ALV 243X and it was operated by B. E. Williams on services between South Wales and Yorkshire. When photographed it was on a return trip approaching the A1 near Ripon (North Yorkshire). *Colin Wright*

MATADOR AND MILITANTS

Still much in evidence during the 1980s were a number of ex-army vehicles dating back to the Second World War (or just after).

AEC 'Matador' USV 833 was seen as a tender carrying diesel for fueling the agricultural vehicles operated by P. J. & I. Foster Crawler Cultivations, of Broadway (in Worcestershire). It was photographed just off the A46 (now the B4632) Cheltenham Road.

Colin Wright

Another AEC 'Matador' USV 110 – this one operated by D. Baldwin & Sons, Keighley, West Yorkshire as a recovery vehicle. Note the registration numbers of these two 4 x 4 AECs USV 833 and USV 110. These are registrations issued by the DVLC for vehicles whose original ones have been 'lost' or were running on trade plates, but where the owner requested a 'period' number.

Colin Wright

RRX 960H was an AEC 'Militant Mk III' operated by Sayers International for use in the recovery of their fleet of maximum weight tankers. Obviously well looked after, as is the rest of the fleet, RRX 960H was photographed at Wroughton Airfield in 1988 on the occasion of the Classic Commercial Rally (organised by the Commercial Vehicle & Road Transport Club). Whilst names are becoming increasingly popular for trucks, it is perhaps not quite so common for them to carry two: 'Wessex Retriever', perhaps its Sunday best, and 'Tiny Tim' – for everyday use? *Colin Wright*

UWT 664N is an older version based on the 'Militant', and the fully slewing crane was provided by Coles Cranes. It was recorded at work on the Great Yorkshire Show Ground at Harrogate, a few miles from its base at Dacre Banks. Owned by E. Houseman & Son, the firm is well known for its international haulage operations.
Colin Wright

CEMENT BULKERS

When this view of two ERF 'C' series cement lorries was taken (in 1987), the name Ribble Cement was almost history, the company in common with Tunnel, Ketton, and Clyde having become part of the Castle Cement empire during the mid-1980s. C579 HCK, a tri-axle unit, was hauling a Metalair trailer still in the attractive grey and white livery of the Clitheroe (North Yorkshire) based Ribble Cement. Hauling a Murfit trailer, the second vehicle in the convoy was registered B491 UBV, a two-axle unit that had been repainted in the equally attractive red and white colours of Castle Cement. The location is Skipton by-pass in North Yorkshire. *Colin Wright*

Working out of Padeswood, C140 JLG was a Foden 'eight-legger' (a type featuring heavily in the cement industry), this one operated by Tunnel Cement. Carrying fleet No. 036/28 it carried the soon to be replaced all-over red and was photographed on the Queensferry by-pass in the summer of 1987. *Colin Wright*

This ERF 'B' series eight-wheeler was some 10 years old when photographed at Evesham. The Rugby Cement orange livery was a familiar sight in the Midlands. *Colin Wright*

The last photograph of this group shows one of Blue Circle Cement's fleet of 32 ton vehicles. Hauled by a Leyland 'Roadtrain' these artics are to be seen nation-wide. C593 FGP was seen at Hunslet, Leeds. *Colin Wright*

HAY AND STRAW

Loads of hay and straw were perhaps one of the most impressive seen on the roads of the 1980s with the obvious exception of abnormal loads. By virtue of its relatively low density, hay and straw can be loaded to maximum dimensions without falling foul of the law and the photographs reproduced give some idea as to the size of these loads.

A brace of Leyland 'Freighter 16-13s' A910 JNB and B123 WJA pose for the camera on the Hartshead Moor Services area of the M62. Both vehicles were operated by J. W. Shaw & Sons, of Saddleworth (on the Yorkshire–Lancashire border). *Colin Wright*

Fir Tree Farm, Grewelthorpe was the home for this 'V' registered Scania '81' operated by J. E. Simpson. TPT 852V and its trailer would gross at about 25 tonnes.
Colin Wright

Carmarthenshire operator Charles Footman's Seddon Atkinson '401' was viewed from the B4293 at the point where the A40 enters the tunnel on its way to Abergavenny and South Wales. Registered MWN 954X this artic carries a neat load of straw. A vehicle of this configuaration could be allowed to gross at 32.5 tonnes, but in this case 22 tonnes would be about the maximum actual weight. *Colin Wright*

Lorries are normally banned from using the 1 in 4 gradient at Park Rash, near Kettlewell (North Yorkshire), but are allowed up if delivering to premises near the road. Dave Norfolk's Scania is on one such trip, hence the rather uneven load. FAT 516Y had arrived with a loaded trailer but it would have been unwise to have attempted to complete the journey with loaded trailer attached. Accordingly this was parked, the '82M' climbing Park Rash to unload, the driver them returning and off-loading his trailer onto the wagon for the final trip. *Colin Wright*

LIVESTOCK TRANSPORT

A quartet of Scanias show an array of axle configuarations in these views of part of the fleet of Cotswold haulier Peter Gilder & Sons. All these vehicles were new to Peter Gilder & Sons and were painted in maroon and white livery. The four views were all taken at the firm's premises near Bourton-on-the-water in Gloucestershire.

The most interesting of the four is undoubtedly the bonneted '142H' registered EFH 294Y hauling a tri-axle semi-trailer built by Parkhouse of Milnthorpe. It was lettered for the Black Bears Polo Team from America and carried their horses.

Peter Gilder/Colin Wright

Our second photograph shows A616 WOK, a three-axle '112M' drawbar unit seen hauling a two-axle trailer. Both wagon and trailer were fitted with bodywork by Bodycraft of Worcester.

Peter Gilder/Colin Wright

'142M' C674 YFH was paired with an unusual trailer with three fixed axles midway along its length built by AHP of Wombourne. Again both bodies are by Bodycraft. *Peter Gilder/Colin Wright*

Our final photograph depicts D760 OOC, another similar '142M' two-axle vehicle but this time hauling a three-axle trailer by M & G (of Lye in the West Midlands). Both bodies are by Parkhouse and are of interest in that they had electricity-operated floors when used as a double-decker. *Peter Gilder/Colin Wright*

PROVEN

Vehicles operated by Craven Arms based Farmore Farmers Ltd are a familiar sight in the English/Welsh border country around Shropshire, Herefordshire and Mid-Wales. Their attractive cream and brown livery is a welcome sight on many farms as they deliver proven (animal feed). The vehicles pictured are:

C494 MAW – the baby of the pack with a gross weight of 7.5 tonnes. This Leyland 'Roadrunner' is fitted with a Boalloy curtain-side body and is used mainly for transferring goods between the company's six depots. *Colin Wright*

Moving up the scale, C270 NAW is depicted in our second photograph. This is a 17 ton gross Leyland 'Freighter' with a Bulkrite body made in nearby Dorrington. Many Farmore Farmers' vehicles carry bodies by this manufacturer. C270 NAW is used mainly on farm deliveries and was photographed on the edge of the company's territory in Staffordshire at the Turf Inn on the A5 at Cannock. *Colin Wright*

A232 YUX is more of a heavyweight with a payload of about 18½ tonnes. This Foden 'S108' eight-wheeler is powered by a Rolls Royce '265' engine and the vehicle is used for the collection of raw materials from the docks on Merseyside, and for bulk farm deliveries of finished goods and grain where the farmer mixes his own compounds. *Colin Wright*

At 38 tonnes maximum gross weight this Volvo 'F10' tri-axle tractor unit C596 MUJ has a lift-up 'tag' axle at the rear. The trailer is based on an M & G chassis, but again Bulkrite produced the bodywork. This rig is used for dockside collection of raw materials on pallets etc., plus farm deliveries of bagged feed and fertiliser. *Colin Wright*

DIVERSITY OF A DALES HAULIER

Ken Longthorne, of Hebden (in the North Yorkshire Dales) has been in the haulage business since 1961 and operated a fleet of smartly turned out tippers and cattle trucks during the 1980s. Now the livestock transport part of the business has been disposed of and the tippers presently carry a blue and maroon livery. In addition to livestock and quarry haulage, Longthorne also had a seasonal job – that of snow-ploughing and gritting. The fleet consisted of eight-wheelers and the odd artic and a handful of six- and four-wheelers.

Volvo 'FL10' tractor with York tri-axle semi-trailer, registered E710 RWY was new to Longthorne in the second half of the 1980s. *Colin Wright*

One of the 'eight-leggers' is shown here in the form of YUA 444X which was purchased second-hand having previously worked for West Yorkshire operater D. N. Smith. It was seen in one of the Yorkshire Dales quarries carrying large boulders for use in flood protection work on the higher reaches of the River Wharfe (for the Yorkshire Water Authority). *Colin Wright*

LJX 269T was one of a pair of DAFs from the late 1970s, a '2300' with bodywork by Houghtons of Milnthorpe. It was photographed in the old cattle market at Skipton which is now the location for Morrisons supermarket. A new cattle market has since been built on the outskirts of the town. The second of the pair of DAFs was LJX 268T – a similar two-axle machine. Both were supplied by Northern Commercials and were regularly seen hauling a drawbar trailer. *Colin Wright*

884 LBP was one of a rare breed – a Thames 'Trader' 6 x 6, one of two in the Longthorne fleet. It was used on a North Yorkshire County Council contract for winter road gritting and snow-ploughing and interestingly could out-perform the local council's Atkinson's 6 x 6 – being lower it had superior traction and on occasions was known to have rescued the aforementioned 'Atki'. 884 LBP was powered by a turbo-charged Ford '360' p.s.v. six-cylinder engine. *Colin Wright*

RURAL PURSUITS

Rural pursuits feature rather strongly in this volume, and this category includes the transport of farm machinery. Vehicles of H. C. Wilson, Elmswell, Suffolk are depicted in the first two photographs, their smart red and white livery being instantly recognisable by the large white 'W' on the cab front. During the 1980s the fleet consisted of eight vehicles which were named to include the word 'Harvester', thus confirming their agricultural role.

'Harvester Prince', F535 NER – a Leyland DAF 'FT595' 6 x 2 and plated to run at 80 tonnes, coupled to a Nooteboom extendable semi-trailer. Its load was a Class '114CS' combine harvester 12ft wide over the wheels. Its operating width was some 18ft hence the cutter bar can be seen placed at right angles on the upper deck of the trailer. *Colin Wright*

A Matrot six-row self-propelled sugar beet harvester measuring some 35ft long x 12ft 6in wide x 13ft high was being transported on a 1983 Tasker 'LB48' tandem-axle trailer in this second view. Power was provided by 'Harvester Duke', registered E182 GBJ – a Scania 'R142M' rigid chassis converted for trailer use and plated at 66 tonnes. *Colin Wright*

LNW 606V was a Scania '111 Intercooler' operated by Cornthwaites, of Pilling, Lancashire. It was seen with a load of second-hand tractors on Hartshead Moor. The four tractors carried are two Massey-Fergusons, a McCormick International, and a Ford.

Colin Wright

E32 RAG was based at Hull with A. S. Haulage Ltd and was observed carrying a load of new Belarus tractors imported from Russia. This 'C' series ERF was powered by a Cummins engine and was supplied by Reliance of Brighouse.

Colin Wright

OUTSIZE LOADS

This group of photographs depict loads that operate outside the normal 'construction and use' regulation but do not come within the romantic scale of heavy haulage as carried out by firms like Pickfords or Wynns. The STGO (Special Types General Order) numbers seen in three of our views perhaps should be explained – there are three categories for different weights, viz: 1 = not exceeding 46 tonnes; 2 = 46–80 tonnes; and 3 = 80–150 tonnes.

The first photograph is an interesting shot taken inside a Yorkshire Dales quarry with Leicester Heavy Haulage Leyland 'S26', registered A614 GNR hauling a mobile stone crusher built by Sheepbridge Engineering. The crusher is powered by a Caterpillar '3412' diesel and is capable of crushing 240 tonnes per hour. The 'load' is being given assistance in the rear by a Komatsu '450' loading shovel, more as insurance against the whole outfit going out of control in reverse.

Colin Wright

Products of the world famous Caterpillar organisation are a familiar sight on our roads with their yellow paintwork. This '235' hydraulic excavator was seen during a rest period being hauled by a Volvo 'F1233' in Cave Plant Hire livery, of Leeds. The trailer rated at 80 tonnes was built by Tasklift, whilst the prime mover was rated at 180 tonnes and with its load a mere 40 tonnes, was child's play for WNW 755X.

Colin Wright

About to traverse the M50, D520 LWN, a DAF '3600' belonging to South Wales operated Lock Contractors Equipment Ltd was seen at Ross-on-Wye. Its load, an NCK crawler crane, was owned by Davies Crane Hire, Carmarthen. These crawler machines were once a familiar sight on building and demolition sites being used not only as cranes but as draglines and with a demolition ball. *Colin Wright*

Heanor Haulage are well known throughout the country wih their red and white fleet of heavy haulage tractors. The company has operated some interesting and unique vehicles in recent years, some of these being produced in their own workshops. However, C351 CYF is a more conventional Volvo 'F12' and was photographed at the A5/A38 roundabout near Lichfield. In this case the load is a pile driver, the horizontal 'arm' at the top swings into the vertical when in use. *Colin Wright*

VEHICLE RECOVERY

A wide selection of makes and types can be seen in the vehicle recovery business. A smart blue and white livery is carried by the quartet of wreckers seen on these two pages – all operated by Teales of Mirfield (West Yorkshire). From 1 January 1988 recovery vehicle licensing changes prohibited the use of trade plates thus leaving the operators with two choices. Generally speaking the age of most wreckers is indeterminate and now 'Q' prefixes, or a 'period' registration, is used.

Q809 HCP is one of a large number of Scammell 'Explorers' that found their way from military to civilian use, this particular vehicle having been built just after World War Two as a 6 x 6 recovery unit. Originally fitted with a Meadows petrol engine it is now powered by a Leyland '680' and is principally used for off-road recovery, its winch will extend for approximately ⅓ mile. Q809 HCP has been in use with Teales for about 10 years. *Colin Wright*

Q810 HCP is slightly older than the vehicle previously described and this Diamond 'T981' 6 x 4 was actually used in the 1939–45 war. When photographed it had been fitted with a Rolls 'C6' engine but other than that and having had recovery equipment added, was more or less in original condition. Its original use was as a ballast box for tank transportation. *Colin Wright*

JSV 843 is a registration of the 'period' category applied to 'Mickey Mouse' eight-wheeler numbered 10 in Teales fleet. Previously this vehicle had been registered 244 BLO in its former life as a sugar tanker with the British Sugar Corporation. This Foden was powered by a Gardner '150'. *Colin Wright*

The fourth of Teales vehicles illustrated – HJX 810N – is much more modern, dating from 1975, but never-the-less had an interesting past whilst operated by James Watkinson of Keighley (West Yorkshire) on heavy haulage duties. A DAF 'F772085DKS', it was converted by Teales from a tractor unit utilising their own crane but with a Holmes A frame. It could lift 8–10 tonnes and had a pull of 108 tonnes. *Colin Wright*

FAIRGROUND

Trucks of the 1980s must inevitably include a handful of old-timers, and where better for a glimpse of times past than the world of the fairground operators.

Our first photograph shows TBT 559H, a Foden 'S36' which travelled widely around Yorkshire and Lincolnshire with Benny Sedgwick's Continental Childrens Ride, the ride itself being a self-contained unit, the build up of which was achieved by lowering the side after removal of the wheels and axles. Foden have, of course, for many years been one of the mainstay manufacturers on the fairground circuit.

Malcolm Slater

ERF has also been popular with the fairground operators and WHH 397 has stood the test of time well, travelling round Scotland with its load of side stalls. This ERF 'KV' was owned by D. Slater and was the last six-wheel 'KV' operating north of the border. The paint job is worthy of note with the ornate lettering and painting of Slater's showmens traction engine superbly rendered. *Malcolm Slater*

The artist really went to town in painting 256 TUR in connection with the owner's haunted castle. The location is Nottingham Goose Fair site in 1982 on the occasion of the annual fair which takes its name from the time when the fair was more of a 'trade' event.

Colin Wright

Another make prevalent on the fairground circuit is Atkinson, in this case one of the six-wheelers that started life as a motorway gritter for the Ministry of Transport. Owned by Herbert Watson Hirst, PYY 974F was seen at Drighlington towing two trailers, a common sight with regard to fairground transport. This Atkinson had a home-made body transferred from an AEC. The frame trailer carries the dodgem cars and the ride roof trusses, whilst the rear trailer is an arcade.

Colin Wright

WASTE DISPOSAL

In recent years the traditional system of waste disposal carried out by local authorities has undergone a major transformation with specialist companies taking over the collection and disposal of the very varied, toxic and dangerous trade refuse. Vehicles belonging to these specialists are featured here with a basic skip lorry in our first photograph.

F960 EKM is one of a vast number of this type of vehicle to be seen nationwide carrying dry non-hazardous waste, the skip having been delivered to the customer's premises for loading and then carted away for disposal of its contents. Greenwich based Quick Skips were the operators of this Leyland 'Freighter' photographed at Erith.

D. G. Savage

Maidstone and Leighton Buzzard based Clean-a-Drain Ltd operated this mammoth gully emptier, based on an ERF 'C' series. D644 KKP was observed in Chelsea Bridge Road, London in January 1987.

D. G. Savage

Two pleasing contrasting blue liveries were carried by the two articulated tankers seen in photographs 3 and 4. In 1985 Volvo introduced its 'FL' range of trucks and an 'FL10' belonging to the Wistech Group (Industrial and Environmental Services) was seen at Ross-on-Wye. The Hazchem code 2X 7017 denoting the carriage of a toxic liquid waste is attached to the Whale tank being hauled by C397 CDE as it passed south-bound on the A1 in Lincolnshire.

Colin Wright

F770 TPU was one of three Cleanaway artics (the other two being Volvo 'F12s') in convoy. Renault 'R365' was hauling a tanker, Hazchem coded 2WE 7008, indicating hazardous waste liquid, containing alkali. A brief explanation of the Hazchem codes is that the four digit number denotes the substance being carried, whilst the figure and letter(s) denote the action to be taken in the event of an accident. Any code with an 'E' means that evacuation of the area around the incident may have to be considered.

Colin Wright

FIRE APPLIANCES

Over the years the country's fire brigades have operated a large number of interesting vehicles, most of which have been based on the smaller end of the available truck models, or on specialised chassis. We present four views:

33 MDT, an AEC 'Mercury' dating from 1963 fitted with a 100ft ladder, was new to Doncaster County Borough Fire Brigade, but passed to South Yorkshire County Fire Service in 1974 under the local government reorganisation. It was photographed at Erskine Road, Rotherham, in 1988 on the occasion of a visit by the Fire Brigade Society. Of the various fire appliances, the heavier chassis fitted with turntable ladders are perhaps the most impressive.

Colin Wright

This newer AEC 'Mercury' registered KWW 228K was fitted with the 'Ergomatic' cab and belonged to the neighbouring West Yorkshire brigade. It was photographed at their Birkenshaw (near Bradford) headquarters. This was one of two such appliances operated by the brigade and had started life with the old West Riding County Fire Service. It was a water carrier with a capacity of 1,400 gallons and was painted white throughout its career which ended in 1986.

Colin Wright

Leyland 'Freighter' D111 TRN was one of three similar appliances based at the East Lancashire town of Accrington. The 'Freighter' replaced Bedford vehicles at about the same time as the sad demise of Bedford as a company to be reckoned with in the fire-engine market, despite a long connection with fire-fighting vehicles. The pump/ladder bodywork on D111 TRN was built by the Scottish company Fulton-Wylie. *Colin Wright*

KPF 814W was operated by the Surrey Fire & Rescue Service and was of Ford manufacture – a make favoured by a number of brigades who purchased 'D' series. Based on a 'D1617' chassis this was one of three in the Surrey brigade. It was built by Cheshire Fire Engineering and carried a Simon 92ft hydraulic platform which was being demonstrated at its home station – Chertsey – to members of the Fire Brigade Society. *Colin Wright*

TELEVISION VEHICLES

Television companies have increased in numbers over the last couple of decades or so, and the industry is now a far cry from the days when the BBC had a virtual monopoly. On these two pages we illustrate four vehicles from four different companies, all used on different jobs.

Nottingham based Chrysalis Television were the operators of Scania '113M', registered E455 BRB. The semi-trailer has bodywork by A. Smith, of Great Bentley, Essex and incorporates a central section which can be extended from 2.6 metres to 3.8 metres in width. Designated 'Unit 3 – Mobile Television Production Unit' it is capable of controlling up to eight cameras and eight videotape recorders.

Geoff Longbottom

A more mudane roll was played by 'Phoenix 2'. This BBC breakdown unit was based on a Bedford 'RL' chassis dating from 1962. It was photographed in 1987 before legislation altered 'trade-plate' law whereby recovery vehicles have to be registered. Based at BBC Manchester this Bedford originally carried registration No. 330 DXP. *Colin Wright*

London-based Visions Mobiles operated this Leyland 'Roadtrain' registered B445 YGN – a mobile television production unit with bodywork by Locomotors (of Andover) on a Richard Froment chassis. The electronic equipment for this unit was supplied and installed by Link Electronics (also of Andover) at a cost of £1 million and was capable of operating up to six cameras and four video-tape recorders. It was later sold to an Australian television company and was replaced by a Leyland 'Freighter' rigid vehicle. *Geoff Longbottom*

A Yorkshire Television 'links vehicle' used to connect an outside broadcast location to the main transmitter network. This example based on a Bedford 4 x 4 illustrated a mid-way point in the transmission chain (in this case on Baildon Moor, near Bradford) employing two roof-mounted parabolic reflector aeriels (dishes), to receive, amplify and retransmit the signal to the next pick-up point. The body was constructed by Lindum Commercial Vehicle Bodies, of Grimsby. *Colin Wright*

COUNCIL VEHICLES

Council vehicles have never been as glamorous as their cousins in the haulage industry, but they are as important and indeed some may say more so. Unfortunately the principal livery of these municipal work-horses is yellow but this at least gives them a high profile in their daily job. In these three photographs we have an interesting selection.

First is a Seddon rear-loading RCV registered HVF 364V operated by Waveney District Council. Waveney was one of the new 'breed' of councils born of the 1974 Local Government reorganisation and covers an area previously centred on Lowestoft, taking in Beccles, Bungay, Southwold, Weirford and Lowestoft. The name 'Waveney' comes from the river of that name which forms the northern boundary of the authority.

Paul Williamson

Our second photograph shows GRW 913N, a Shelvoke & Drewry (a make synonymous with council duties since the early days of the present century), operated by Stratford-on-Avon District Council. *Paul Williamson*

Wales features in the last of the trio of council vehicles with the location near the Heads of the Valleys road (A465) close to Merthyr Tudfil. The lettering on the cab door of this Leyland 'Freighter 16-17' tipper registered B172 BAX, reads 'Cyngor Bwrdeisdref, Merthyr Tudful'. It is interesting to note that the Leyland 'Freighter' claimed a proportion of the market lost to General Motors when the Bedford range was discontinued.

Colin Wright